AF270696

BOBCATS

by Elizabeth Andrews

Cody Koala

An Imprint of Pop!
popbooksonline.com

abdobooks.com

Published by Pop!, a division of ABDO, PO Box 398166, Minneapolis, Minnesota 55439. Copyright ©2023 by Abdo Consulting Group, Inc. International copyrights reserved in all countries. No part of this book may be reproduced in any form without written permission from the publisher. Cody Koala™ is a trademark and logo of Pop!.

Printed in the United States of America, North Mankato, Minnesota

052022
092022

Cover Photo: Shutterstock Images
Interior Photos: Shutterstock Images

Editor: Grace Hansen
Series Designer: Laura Graphenteen

Library of Congress Control Number: 2021951844
Publisher's Cataloging-in-Publication Data
Names: Andrews, Elizabeth, author.
Title: Bobcats / by Elizabeth Andrews
Description: Minneapolis, Minnesota : Pop, 2023 | Series: Twilight Animals | Includes online resources and index
Identifiers: ISBN 9781098242077 (lib. bdg.) | ISBN 9781098242770 (ebook)
Subjects: LCSH: Bobcat--Juvenile literature. | Bobcat--Behavior--Juvenile literature. | Twilight--Juvenile literature. | Nocturnal animals--Juvenile literature. | Nocturnal animals--Behavior--Juvenile literature.
Classification: DDC 591.518--dc23

Hello! My name is

Cody Koala

Pop open this book and you'll find QR codes like this one, loaded with information, so you can learn even more!

Scan this code* and others like it while you read,

or visit the website below to make this book pop.

popbooksonline.com/bobcats

*Scanning QR codes requires a web-enabled smart device with a QR code reader app and a camera.

Table of Contents

Chapter 1

Sneaky Cats

Moving slowly and silently through the underbrush, a bobcat hunts. It is nearly invisible against the fallen leaves on the dark forest floor. The **prey** the bobcat hunts is most active at twilight.

Watch a video here!

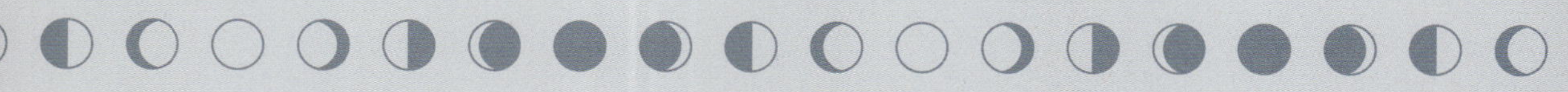

The coat of a bobcat can be tan to brown with a reddish **tinge**. The coat also has dark spots and stripes. Bobcats are perfectly **camouflaged**, especially when night falls.

Bobcats are sometimes called wildcats.

Chapter 2

Made to Hunt

It is not just the bobcat's coat that helps it hunt. Its ears are important and have extra fur to help the big cat hear better. But a bobcat's most important sense when hunting is its eyesight.

Learn more here!

spotted coat
good for camouflage

ears
extra furry to
capture sound

eyes
strong eyesight
to hunt prey

paw
large pads to
walk quietly

Bobcats have large paws
so they can walk silently.
Their legs are long and help
them run fast and pounce
on their prey. Bobcats have
sharp claws. Their claws
are used for hunting and
climbing trees.

Bobcats will eat anything.
Usually they hunt rabbits,
rodents, birds, and fish. In
many areas, bobcats hunt

and eat deer. They play an important role in keeping deer **populations** from getting too big.

Chapter 3

Territorial

Bobcats are **solitary** creatures. This means they live alone most of the time. They will walk about seven miles (11.3km) of their territory every night to hunt.

Demonstrate or show a video of your experiment.

Create comics or other drawings to show your project in a fun way.

Include props, models, or dioramas.

One way to present a project is with a display board. It should show how you followed the scientific method. Turn the page to see a display board of the project in this book!

FRICTION & MOTION

QUESTION

How does friction affect the speed of a moving object?

RESEARCH

Friction is a force that acts against a moving object. Friction occurs between two objects or surfaces that are touching each other. Rough surfaces create more friction than smooth surfaces.

HYPOTHESIS

I think an increase in friction results in a decrease in speed.

EXPERIMENT

The purpose of my experiment was to learn how friction affects motion. In my experiment, I tested how fast a toy car moved down tracks of varying roughness.

RESULTS

My results showed that the rougher the track material, the longer it took the car to reach the bottom of the track.

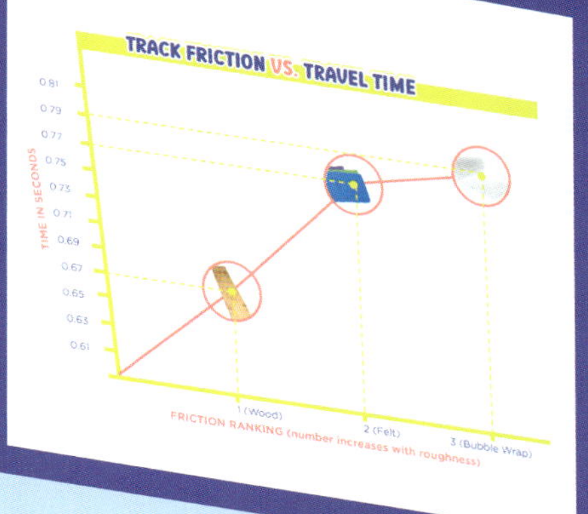

CONCLUSION

The results support my hypothesis that an increase in friction results in a decrease in speed.

KEEP ASKING QUESTIONS

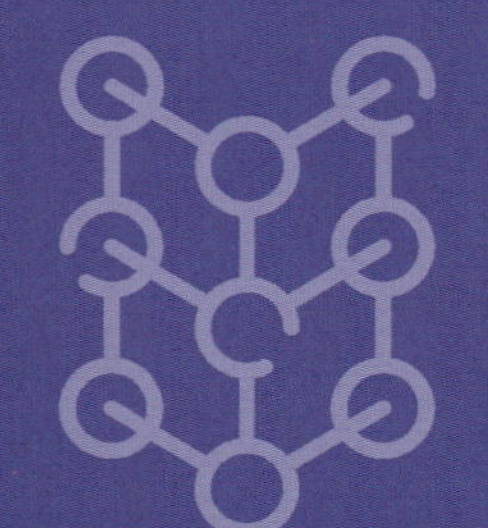

Your science project is over. You packed away your display. But don't stop asking questions! What might you do differently if you did the project again? What additional **research** could you do? Is there a related **topic** you would like to explore?

BEYOND THE SCIENCE FAIR

Be a scientist beyond the science fair! You can use parts of the scientific method to find answers to everyday questions. Maybe you have a hypothesis for why you move down the slide faster than your mom. Maybe you experiment with ways to prevent people from slipping on icy sidewalks. One day, you might use science to do big things. Maybe you'll **design** superfast trains! Turn your world into a science fair. What will you discover?

GLOSSARY

bias—showing a preference for one result over another.

calculate—to use math to figure something out.

design—to plan how something will appear or work.

focus—to concentrate on or pay particular attention to.

friction—a force that slows motion when two surfaces touch each other.

future—the time that hasn't happened yet.

information—the facts known about an event or subject.

matter—anything that has weight and takes up space.

research—to find out more about something. Also, a study of something to learn new information.

summary—a short statement of the main points.

topic—the main idea or subject.

variable—a factor in a scientific experiment that may change.

Bobcats have short tails that look like they have been **bobbed**. That is how they got their name!

Male bobcats usually control a large area of land. They will allow females to stay on small parts of it, but never other male bobcats. Territory is marked with scent and claw marks.

Chapter 4

Babies

Bobcats come together for a few days each year to have young. The female bobcat cares for her kittens on her own. The family uses a **cavity**, like a hollow tree stump, for a den.

Complete an
activity here!

Kittens drink their mother's
milk for two months. They
will stay with their mother
no more than a year.

During that time, bobcat kittens learn to hunt and survive in the wild.

The white marks on bobcats' ears and tails help cubs find their moms in the dark.

Making Connections

Text-to-Self

Do you think you could catch sight of a bobcat? How would you try to find it?

Text-to-Text

Have you read about different big cats in other books? How were they similar to or different from the bobcat?

Text-to-World

Bobcats' camouflage helps them hunt. What are some other animals with good camouflage?

Glossary

bob – to cut shorter.

camouflage – having colors or patterns that blend into surroundings.

cavity – unfilled space.

population – the number of beings in an area.

prey – an animal that is hunted.

solitary – living alone.

tinge – a small amount.

Index

Online Resources

popbooksonline.com

Thanks for reading this Cody Koala book!

Scan this code* and others like it in this book, or visit the website below to make this book pop!

popbooksonline.com/bobcats

*Scanning QR codes requires a web-enabled smart device with a QR code reader app and a camera.